AF409358

There is not any event horizon

CARLO MARIA PACE

THERE IS NOT ANY EVENT HORIZON

Youcanprint

Title | **THERE IS NOT ANY EVENT HORIZON**
Author | CARLO MARIA PACE

ISBN | 978-88-31645-99-7

© 2019 - All rights reserved to the Author
This work is published directly by the Author through the Youcanprint self-publishing platform and the Author holds every right of the same in an exclusive manner. No part of this book can therefore be reproduced without the prior consent of the author.

Youcanprint
Via Marco Biagi 6 - 73100 Lecce
www.youcanprint.it
info@youcanprint.it

INTRODUCTION

Commonly, for the determination of the value of the gravitational mass of a particle (or of a body) all the contributions to the energy of the particle (or of the body) are summed together except the gravitational potential energy, that is except the gravitational binding energy. For example, the nuclear binding energies of the nucleons in the atomic nuclei contribute directly to the usual determination of the gravitational masses of the atomic nuclei: for this reason, the mass of each atomic nucleus is considered to be less than the sum of the masses of the nucleons of the same atomic nucleus. But, in the usual treatment of a particle (or of a body) in a gravitational field, the contribution of the gravitational potential energy, that is of the gravitational binding energy, to the determination of the gravitational mass of the particle (or of the body) in the gravitational field is not taken into account in any way. This is a profoundly incorrect treatment; in particular, when the gravitational field is very intense the results are very different from those of the correct treatment.

Therefore, in the first place, we, by considering also the contribution of the gravitational potential energy to the determination of the gravitational mass of a particle (or of a small mass), intend to find the correct formulas of the gravitational potential energy and of the total energy of a particle (or of a small mass) in a gravitational field due to a spherically symmetric static mass.

CHAPTER I
THE DYNAMICS
IN ANY GRAVITATIONAL FIELD
DUE TO A SPHERICALLY
SYMMETRIC STATIC MASS

1 The motion of a particle (or of a small mass) in the case of a gravitational field due to a spherically symmetric static mass

In the first place, we examine the dynamics of a particle (or of a small mass) B in the case of a gravitational field due to a spherically symmetric static mass M. We will use an inertial reference frame that is integral with the mass M. Moreover, we will use the centre of the mass M as origin of this reference frame. For simplicity we will consider negligible the perturbation produced by the small mass of B on the gravitational field due to the central mass M. Moreover, we will consider negligible the possible irradiation from the same particle (or from the same small mass) B because of its acceleration.

A motionless particle (or a motionless small mass) B, that has a rest mass $m_0 \ll M$, at $r = +\infty$, where r is the distance of B from the centre of the mass M (for simplicity, in this book, we will consider that B is always outside the mass M for all the values of r greater than zero), has a gravitational potential energy equal to zero. If this particle (or small mass) B moves because of the effect of the gravitational field in the direction of the central mass M, the same particle (or small mass) B, for the principle of conservation of energy, acquires a kinetic energy equal to the decrease of its gravitational potential energy that becomes negative. Therefore, during the motion of B, the total energy of B always remains the same and consequently, for the two Einsteinian theories of Relativity[1], both the inertial mass of B and the gravitational mass of B (which masses, moreover, for the principle of equivalence, are always equal to each other) always remain the same. In particular, for $r \to 0$ we have that the kinetic energy of B tends to $+\infty$ according to the relativistic formulas (of course, without that B exceeds the velocity of the light in vacuum c relatively to any inertial reference frame) while the gravitational potential energy of B tends to $-\infty$, in such a way as to maintain constant the total energy of B and therefore (to maintain constant) also both the gravitational mass of B and the inertial mass of B (which masses are always equal to each other). Obviously, in any moment of this motion of B, if we reverse the direction of the motion of B, B would return to its initial motionless state at $r = +\infty$. In particular, the particle (or small mass) B, during the return back, will decrease its kinetic energy, always in such a way that, for the principle of conservation of energy and the principle of equivalence, its inertial mass and its gravitational mass always remain the same during the motion (also here, moreover, these two masses, for the principle of equivalence, are always equal to each other).

More precisely, for the principle of equivalence, we always have:

$$m_g = m_i \tag{1}$$

---

[1] Cf. EINSTEIN (1905); EINSTEIN (1916).

where m_g is the gravitational mass of B, and m_i is the inertial mass of B.

Moreover, for the Special Theory of Relativity, we have:

$$m_i = \frac{E_T}{c^2} \qquad (2)$$

where E_T is the total energy of B, and c is the velocity of light in vacuum.

Therefore from the (1) and the (2) we have:

$$m_g = \frac{E_T}{c^2} \qquad (3)$$

Now, by using the (3), we have:

$$E_g = P_g(r)m_g = P_g(r)\frac{E_T}{c^2} \qquad (4)$$

where E_g is the gravitational potential energy of B, and $P_g(r)$ is the gravitational potential (that is the gravitational potential energy per unit of mass) of B. In the (4) we have written P_g as a function only of the variable r, that is as $P_g(r)$, because we are considering a gravitational field due to a spherically symmetric central static mass: in this case the gravitational field is independent of the direction and therefore the gravitational potential can be dependent only on the variable r. Moreover, in this case, the total energy E_T of B is the sum of the rest energy of B, of the kinetic energy of B and of the gravitational potential energy of B, for which we have:

$$E_T = m_0 c^2 + m_0 c^2 (\gamma_r - 1) + P_g(r) \frac{E_T}{c^2} \qquad (5)$$

where $m_0 c^2$ is the rest energy of B, $m_0 c^2 (\gamma_r - 1)$ is the kinetic energy of B at r, and $\gamma_r \equiv \dfrac{1}{\sqrt{1 - \frac{v_r^2}{c^2}}}$, where v_r is the velocity of B at r.

On the other hand, we have noted that the total energy of B is always the same; for which we also have:

$$E_T = m_0 c^2 \qquad (6)$$

Now, by substituting the (6) in the (5), we have:

$$m_0 c^2 = m_0 c^2 + m_0 c^2 (\gamma_r - 1) +$$

$$+ P_g(r) \frac{m_0 c^2}{c^2} \qquad (7)$$

From the (7) we have:

$$0 = m_0 c^2 (\gamma_r - 1) + P_g(r) m_0 \qquad (8)$$

From the (8), finally, we have:

$$P_g(r) = -c^2(\gamma_r - 1) \tag{9}$$

Moreover, from the (4) and the (6) the gravitational potential energy is equal to:

$$E_g = P_g(r)m_0 \tag{10}$$

Now, if we subtract the kinetic energy to B, the expression of E_T, that in this case is equal to the sum only of the rest energy of B and of the gravitational potential energy of B, becomes:

$$E_T = m_0 c^2 + P_g(r)\frac{E_T}{c^2} \tag{11}$$

From the (11) we have:

$$E_T\left[1 - \frac{P_g(r)}{c^2}\right] = m_0 c^2 \tag{12}$$

Now, by using the (9), the (12) becomes:

$$E_T\left[1 - \frac{-c^2(\gamma_r - 1)}{c^2}\right] = m_0 c^2 \tag{13}$$

From the (13) we have:

$$E_T[1 + (\gamma_r - 1)] = m_0 c^2 \tag{14}$$

And, finally, from the (14) we have:

$$E_T = \frac{m_0 c^2}{\gamma_r} \tag{15}$$

Consequently, from the (4) and the (15), the gravitational potential energy in this case is equal to:

$$E_g = P_g(r)\frac{E_T}{c^2} = P_g(r)\frac{m_0}{\gamma_r} \tag{16}$$

Therefore, this particle (or small mass) B can go from r up to $+\infty$ if we give to it a kinetic energy equal or superior to

$$m_0 c^2(\gamma_r - 1) = -P_g(r)m_0 \tag{17}$$

In other words, B can go from r up to $+\infty$ if and only if B has a velocity equal or superior to v_r. Obviously, the kinetic energy of the formula (17) is the necessary contribution for increasing the total energy of the particle (or small mass) to the value of energy of a stationary free particle (or small mass), that is to $m_0 c^2$. In fact, if we gave to B a kinetic energy equal to that of the expression (17), we would return to the situation described by the (5) and the (6).

In real situations the particle (or small mass) B can irradiate and can interact with other particles, and therefore can decrease or can increase in its total energy: but, in any case, its total energy cannot decrease to negative values and if its total energy becomes zero we have in reality that B by means of its interactions has been transformed in something else.

More in general, if we do not start with a rest particle (or small mass) at $r = +\infty$, we have that the γ of the particle (or small mass) is independent of the γ_r that figures in the expression of the gravitational potential. Therefore, in general the (5) becomes:

$$E_T = m_0 c^2 + m_0 c^2 (\gamma - 1) + P_g(r) \frac{E_T}{c^2} \qquad (18)$$

From which we have:

$$E_T \left[1 - \frac{P_g(r)}{c^2} \right] = m_0 c^2 \gamma \qquad (19)$$

Now, also in this case the (9) is valid. Therefore the (19), by using the (9), becomes:

$$E_T \left[1 - \frac{- c^2 (\gamma_r - 1)}{c^2} \right] = m_0 c^2 \gamma \qquad (20)$$

From which we have:

$$E_T \left[1 + (\gamma_r - 1) \right] = m_0 c^2 \gamma \qquad (21)$$

From which, finally, we have:

$$E_T = \frac{m_0 c^2 \gamma}{\gamma_r} \qquad (22)$$

This expression is the generalization of the (6), in which $\gamma = \gamma_r$, and of the (15), in which $\gamma = 1$.

This expression (22) shows that the particle (or small mass) is free to go up to $r = +\infty$ if and only if $\gamma \geq \gamma_r$. Since γ_r is finite and for $v \to c$, $\gamma \to +\infty$, there is always an escape velocity, less than c, for the particle (or small mass) B. In particular this escape velocity is equal to v_r, which is the velocity that figures in the expression of γ_r.

Moreover in this general case, by using the (4) and the (22), we have:

$$E_g = P_g(r)\frac{E_T}{c^2} = P_g(r)\frac{m_0\gamma}{\gamma_r} \qquad (23)$$

This expression is the generalization of the (10), in which $\gamma = \gamma_r$, and of the (16), in which $\gamma = 1$.

Moreover, by using the (3) and the (22), we have that in this case the gravitational mass of B is equal to:

$$m_g = \frac{E_T}{c^2} = m_0\frac{\gamma}{\gamma_r} \qquad (24)$$

On the other hand, by using the usual incorrect definition of the gravitational mass we would have, instead of the (5) and of the (7), respectively these equations:

$$E_T = m_0c^2 + m_0c^2(\gamma_r - 1) +$$

$$+ P_g(r)\frac{m_0c^2\gamma_r}{c^2} \tag{25}$$

$$m_0c^2 = m_0c^2 + m_0c^2(\gamma_r - 1) +$$

$$+ P_g(r)\frac{m_0c^2\gamma_r}{c^2} \tag{26}$$

From the (26) we have:

$$P_g(r) = -c^2\frac{(\gamma_r - 1)}{\gamma_r} = -c^2\left(1 - \frac{1}{\gamma_r}\right) \tag{27}$$

Obviously, the (27) is the analogue of the (9). Now, because γ_r by definition can be only ≥ 1, according to the (27) $P_g(r)$ could be only in an interval between $-c^2$ and 0, that is in this case we would have:

$$-c^2 < P_g(r) \leq 0 \tag{28}$$

In particular, from the (27) we would have that $P_g(r) = 0$ for $\gamma_r = 1$, and $P_g(r) \rightarrow -c^2$ for $\gamma_r \rightarrow +\infty$. Now we can reasonably suppose that $P_g(r)$ must be directly proportional to M, for which the (28) cannot always be valid if not only in the case in which $P_g(r)$ is identically null; but, on the other hand, $P_g(r)$ is identically null only when there is not any gravitational field. Consequently, the (28) and therefore the (27) are completely unacceptable. Moreover, because the (27) and the (28) are a direct consequence of the (26), we must consider invalid the (26). But, if we consider valid the erroneous formula (25), we can consider invalid the (26) only by affirming that the principle of conservation of energy in this case is not valid, but also this way of proceeding would obviously be unacceptable, since the principle of conservation of energy is incontestable in physics. In other words, the erroneous formula (25) implies the unacceptable violation of the principle of conservation of energy. In conclusion we must use the correct definition of the gravitational mass, that is we must use the (3), in which we take into account also the contribution of the gravitational potential energy to the determination of the gravitational mass.

2 Extension to the case of a photon

On the other hand, if we consider a photon, instead of a particle (or small mass) B with rest mass that is different from zero, in the case that we suppose that the photons have a very small rest mass that is consistent with the experimental results, then we can proceed in the same way.

On the contrary, in the case that we suppose that the rest mass of the photons is rigorously equal to zero, then the total energy hf_0 of a photon at $r = +\infty$ corresponds to the rest energy $m_0 c^2$ of the particle (or small mass) B. Moreover the photon instead of increasing its velocity increases its frequency during its motion towards the central mass M in such a way that, for the principle of conservation of energy and the principle of equivalence, its inertial mass and its gravitational mass always remain the same during the motion (which masses, moreover, for the principle of equivalence, are always equal to each other). Analogously, as for the possible return back of the photon up to $r = +\infty$, the photon, during the return back, will decrease its frequency instead of decreasing its velocity, always in such a way that, for the principle of conservation of energy and the principle of equivalence, its inertial mass and its gravitational mass always remain the same during the motion (also here, moreover, these two masses, for the principle of equivalence, are always equal to each other). On the other hand, this reasoning is obviously valid not only for the photons but also, in an analogous way, for all the other particles that are considered to have rest mass equal to zero.

More precisely, instead of the sum of the rest energy and the kinetic energy a photon has an energy equal to:

$$E = \frac{hc}{\lambda} = hf \tag{29}$$

where h is the Planck constant, λ is the wavelength of the photon and f is the frequency of the photon. Obviously hf is the analogue in the case of the photon of $m_0 c^2 \gamma$ in the case of a particle with rest mass equal to m_0.

Therefore the total energy in this case, by using the (4) and the (29), is:

$$E_T = hf + P_g(r)\frac{E_T}{c^2} \tag{30}$$

From which we have:

$$E_T\left[1 - \frac{P_g(r)}{c^2}\right] = hf \tag{31}$$

From the (31), by using the (9), we have:

$$E_T\left[1 - \frac{-c^2(\gamma_r - 1)}{c^2}\right] = hf \qquad (32)$$

From which we have:

$$E_T[1 + (\gamma_r - 1)] = hf \qquad (33)$$

From which, finally, we have:

$$E_T = \frac{hf}{\gamma_r} \qquad (34)$$

This is the analogue, in the case of a photon, of the formula (22) and, in the case of $f = f_0$, of the formula (15). Here, similarly to the case of a particle with a rest mass greater than zero, we can give energy to the photon by multiplying its frequency by a factor γ_r and so the photon would have the energy for freeing itself from the gravitational field with a frequency f.

On the other hand, as for a photon in a situation analogous to that of B without kinetic energy in the gravitational field, that is in the situation (of B) described by the equation (15), the photon (always in the hypothesis that the rest mass of a photon is rigorously equal to zero) can always go up to $r = +\infty$, because in that case the frequency of the photon would be simply reduced by the factor γ_r that figures in the (34). In other words, a photon (always in the hypothesis that the rest mass of a photon is rigorously equal to zero) is always free to go up to $r = +\infty$, since the energy of the photon in a gravitational field, according to the (34), is always positive and its rest mass is rigorously zero: in fact a photon, if its rest mass is rigorously zero, can have at $r = +\infty$ an arbitrarily small energy and therefore a photon, that always has a positive energy in a gravitational field according to the (34), always has the sufficient energy for arriving at $r = +\infty$.

In this case, from the (3) and the (34) we have that the gravitational mass is equal to:

$$m_g = \frac{E_T}{c^2} = \frac{hf}{c^2 \gamma_r} \tag{35}$$

On the other hand, by using the usual incorrect definition of the gravitational mass we would have, instead of the (30), this equation:

$$E_T = hf + P_g(r)\frac{hf}{c^2} \tag{36}$$

And, instead of the (7), by taking into account that in this case hf is the analogue of $m_0 c^2 \gamma_r$, we would have:

$$\frac{hf}{\gamma_r} = hf + P_g(r)\frac{hf}{c^2} \tag{37}$$

From the (37) also in this case we would have:

$$P_g(r) = -c^2 \frac{(\gamma_r - 1)}{\gamma_r} = -c^2\left(1 - \frac{1}{\gamma_r}\right) \qquad (38)$$

Obviously, the (38) is equal to the (27) and, as the (27), is the analogue of the (9). Now, also here, because γ_r by definition can be only ≥ 1, according to the (38) $P_g(r)$ could be only in an interval between $-c^2$ and 0, that is in this case we would have:

$$-c^2 < P_g(r) \leq 0 \qquad (39)$$

In particular, from the (38) we would have that $P_g(r) = 0$ for $\gamma_r = 1$, and $P_g(r) \to -c^2$ for $\gamma_r \to +\infty$. Now we can reasonably suppose that $P_g(r)$ must be directly proportional to M, for which the (39) cannot always be valid if not only in the case in which $P_g(r)$ is identically null; but, on the other hand, $P_g(r)$ is identically null only when there is not any gravitational field. Consequently, the (39) and therefore the (38) are completely unacceptable. Moreover, because the (38) and the (39) are a direct consequence of the (37), we must consider invalid the (37). But, if we consider valid the erroneous formula (36), we can consider invalid the (37) only by affirming that the principle of conservation of energy in this case is not valid, but also this way of proceeding would obviously be unacceptable, since the principle of conservation of energy is incontestable in physics. In other words, the erroneous formula (36) implies the unacceptable violation of the principle of conservation of energy. In conclusion, also in the case of a photon, we must use the correct definition of the gravitational mass, that is we must use the (3), in which we take into account also the contribution of the gravitational potential energy to the determination of the gravitational mass.

CHAPTER II
THE DYNAMICS IN THE CASE OF A SEMI-NEWTONIAN APPROXIMATION

1 A semi-Newtonian approximation

Up to this point, the treatment that we have developed is completely general and therefore we can affirm that this treatment must be valid for any theory of gravitation, in particular for the General Theory of Relativity and for any other theory even more correct of the gravitation, provided that the principle of conservation of energy, the Special Theory of Relativity and the principle of equivalence are correct.

Now, we make a semi-Newtonian approximation, that is, to the principle of conservation of energy, the Special Theory of Relativity and the principle of equivalence, we also add the Newton's law of universal gravitation:

$$F_g = \frac{GM}{r^2} m_g \qquad (40)$$

and the Newtonian formula for the gravitational potential:

$$P_g(r) = -\frac{GM}{r} \qquad (41)$$

Therefore, by using the (9) and the (41) we have:

$$-c^2(\gamma_r - 1) = -\frac{GM}{r} \tag{42}$$

From which we have:

$$(\gamma_r - 1) = \frac{GM}{rc^2} \tag{43}$$

From which, finally, we have:

$$\gamma_r = 1 + \frac{GM}{rc^2} \tag{44}$$

Therefore the expression (23) for the gravitational potential energy, by using the (44) and the (41), becomes:

$$E_g = -\frac{GM}{r} m_0 \gamma \frac{1}{1 + \dfrac{GM}{rc^2}} =$$

$$= -\frac{GM}{rc^2} m_0 c^2 \gamma \frac{1}{1 + \dfrac{GM}{rc^2}} \tag{45}$$

On the other hand, the expression (22) for the total energy, by using the (44), becomes:

$$E_T = m_0 c^2 \gamma \, \frac{1}{1 + \dfrac{GM}{rc^2}} \tag{46}$$

On the other hand, in this case the (15) and the (16) become respectively the (46) and the (45) with $\gamma = 1$, that is we have:

$$E_T = m_0 c^2 \, \frac{1}{1 + \dfrac{GM}{rc^2}} \tag{47}$$

$$E_g = -\frac{GM m_0}{r} \, \frac{1}{1 + \dfrac{GM}{rc^2}} \tag{48}$$

Moreover, in this case, by using the (3) and the (46), we have that the gravitational mass is equal to:

$$m_g = \frac{E_T}{c^2} = m_0 \gamma \, \frac{1}{1 + \dfrac{GM}{rc^2}} \tag{49}$$

From the equation (46) we see that a particle (or small mass) can free itself from the gravitational field if and only if its velocity is such that we have: $\gamma \geq 1 + \dfrac{GM}{rc^2}$.

Now, since M is always finite and r is always greater than zero, the value of γ that is equal to $1 + \dfrac{GM}{rc^2}$ is always a finite value. Therefore, there is always an escape velocity v_e less than the velocity of light in vacuum: the escape velocity v_e is obviously the value of velocity for which γ is equal to $1 + \dfrac{GM}{rc^2}$, that is we have:

$$\frac{1}{\sqrt{1 - \dfrac{v_e^2}{c^2}}} = 1 + \frac{GM}{rc^2} \qquad (50)$$

From the (50) we have:

$$\frac{1}{1 - \dfrac{v_e^2}{c^2}} = \left(1 + \frac{GM}{rc^2}\right)^2 \qquad (51)$$

From which we have:

$$1 = \left(1 + \frac{GM}{rc^2}\right)^2 \left(1 - \frac{v_e^2}{c^2}\right) \qquad (52)$$

From which we have:

$$\frac{v_e^2}{c^2}\left(1 + \frac{GM}{rc^2}\right)^2 = \left(1 + \frac{GM}{rc^2}\right)^2 - 1 \qquad (53)$$

From which, finally, we have:

$$\frac{v_e^2}{c^2} = 1 - \frac{1}{\left(1 + \frac{GM}{rc^2}\right)^2} \tag{54}$$

On the other hand, the erroneous formula currently used instead of the (46) is:

$$E_T = m_0 c^2 + m_0 c^2 (\gamma - 1) - \frac{GM}{rc^2} m_0 c^2 \gamma =$$

$$= m_0 c^2 \gamma \left(1 - \frac{GM}{rc^2}\right) \tag{55}$$

We can note that in the (55) the gravitational mass is erroneously considered to be equal to the sum only of the rest mass and of the contribution to the mass by the kinetic energy, because the contribution of the gravitational potential energy to the determination of the gravitational mass is erroneously considered to be null.

According to this erroneous formula (55), if $\frac{GM}{rc^2} \geq 1$, the total energy E_T would be negative or null for all values of γ, and therefore the total energy E_T would be negative or null for all values of the velocity of the particle (or small mass). Therefore, according to the erroneous formula (55), there would not be any escape velocity for the particle (or small mass) for values of r less than or equal to $\frac{GM}{c^2}$. Moreover, according to the erroneous formula (55) there would not be the conservation of energy in the motion of a particle (or small mass) in the gravitational field from $r = +\infty$ to a value of r less than $\dfrac{GM}{c^2}$, because at $r = +\infty$ the total energy of the particle (or small mass) is $\geq m_0 c^2$, while for values of r less than $\dfrac{GM}{c^2}$ the total energy of the particle (or small mass), according to the (55), would be negative. In other words, the erroneous formula (55) violates the principle of conservation of energy.

Obviously, the correct formula (46) and the incorrect formula (55) are practically equal to each other for the values of M and r for which we have: $\dfrac{GM}{rc^2} \ll 1$, that is when we can neglect $\dfrac{GM}{rc^2}$ with respect to 1.

As for the case of a photon, the formula (34), by using the (44), becomes:

$$E_T = \frac{hf}{\gamma_r} = hf \frac{1}{1 + \dfrac{GM}{rc^2}} \tag{56}$$

Obviously, also in this case a photon is always free to go up to $r = +\infty$. In fact, in this case the frequency of the photon at $r = +\infty$ becomes:

$$f_0 = \frac{f}{\gamma_r} = f \frac{1}{1 + \dfrac{GM}{rc^2}} \tag{57}$$

Now, we have that f_0 is always less than f, but we also have that f_0 is always greater than zero. Therefore the photon (if its rest mass is rigorously equal to zero) always has the energy for going up to $r = +\infty$.

Finally, by considering that, in the case of a photon, hf is the analogue of $m_0 c^2 \gamma$ in the case of a particle with rest mass greater than zero, the erroneous formula (55) in the case of a photon becomes:

$$E_T = hf - \frac{GM}{rc^2} hf = hf \left(1 - \frac{GM}{rc^2} \right) \tag{58}$$

Also here, we can note that in the (58) the gravitational mass is erroneously considered to be equal to the sum only of the rest mass and of the contribution to the mass by the kinetic energy, because the contribution of the gravitational potential energy to the determination of the gravitational mass is erroneously considered to be null.

According to this erroneous formula (58), if $\frac{GM}{rc^2} \geq 1$, the total energy E_T would be negative or null for all values of f, and therefore the total energy E_T would be negative or null for all values of the frequency of the photon. Therefore, according to the erroneous formula (58), for values of r less than or equal to $\frac{GM}{c^2}$ there would not be any escape frequency for a photon, that is, a photon with whatever frequency could never free itself from the gravitational field. Moreover, also here, according to the erroneous formula (58) there would not be the conservation of energy in the motion of a photon in the gravitational field from $r = +\infty$ up to a value of r less than $\frac{GM}{c^2}$, because at $r = +\infty$ the total energy of the photon is $hf_0 > 0$, while for values of r less than $\frac{GM}{c^2}$ the total energy of the photon, according to the (58), would be negative. In other words, the erroneous formula (58) violates the principle of conservation of energy.

Obviously, also here, the correct formula (56) and the incorrect formula (58) are practically equal to each other for the values of M and r for which we have: $\dfrac{GM}{rc^2} \ll 1$, that is when we can neglect $\dfrac{GM}{rc^2}$ with respect to 1.

2 The stable circular orbital motions in the semi-Newtonian approximation

Moreover, in this case the gravitational force due to M on the particle (or small mass) B, by using the (40) and the (49), is a centripetal force equal to:

$$F_g = \frac{GM}{r^2} m_0 \gamma \frac{1}{1 + \dfrac{GM}{rc^2}} \tag{59}$$

On the other hand, for having a stable circular orbital motion with a radius r the centripetal force, by taking into account the Special Theory of Relativity, has to be equal to:

$$m_0 \gamma \frac{v^2}{r} \tag{60}$$

Therefore we have a stable circular orbital motion of radius r when the (59) is equal to the (60), that is when we have:

$$\frac{GM}{r^2} m_0\gamma \frac{1}{1 + \dfrac{GM}{rc^2}} = m_0\gamma \frac{v^2}{r} \tag{61}$$

From which we have:

$$\frac{GM}{r} \frac{1}{1 + \dfrac{GM}{rc^2}} = v^2 \tag{62}$$

From which we have:

$$\frac{GM}{rc^2} \frac{1}{1 + \dfrac{GM}{rc^2}} = \frac{v^2}{c^2} \tag{63}$$

From which we have:

$$\frac{\dfrac{GM}{rc^2} + 1 - 1}{1 + \dfrac{GM}{rc^2}} = \frac{v^2}{c^2} \tag{64}$$

From which, finally, we have:

$$1 - \frac{1}{1 + \dfrac{GM}{rc^2}} = \frac{v^2}{c^2} \tag{65}$$

The first member of the (65), for every value of r greater than zero, is always positive and less than 1, and therefore for every value of r greater than zero there is always a value of v less than c for which value there is a stable circular orbital motion of radius r.

In particular, we have that for $r \rightarrow +\infty$, the first member of the (65) tends to zero; and therefore for the equation (65) we have that for $r \rightarrow +\infty$, $v \rightarrow 0$.

Moreover, we have that for $r \rightarrow 0$, the first member of the (65) tends to 1; and therefore for the equation (65) we have that for $r \rightarrow 0$, $v \rightarrow c$.

All this is consistent with what we have seen, that is with the fact that a photon is always free to go up to $+\infty$ and that the escape velocity at every r is always less than c. Moreover, from the (54) and the (65) we can note that, correctly, for every finite value of r greater than zero, the escape velocity (from r) is always greater than the velocity for which we have a stable circular orbital motion of radius r.

On the other hand, the erroneous formula commonly used instead of the (59) is:

$$F_g = \frac{GM}{r^2} m_0 \gamma \tag{66}$$

Consequently, the erroneous equation currently used instead of the (61) is:

$$\frac{GM}{r^2} m_0 \gamma = m_0 \gamma \frac{v^2}{r} \tag{67}$$

From the (67) we have that the erroneous formula currently used instead of the (65) is:

$$\frac{GM}{rc^2} = \frac{v^2}{c^2} \qquad (68)$$

We can note that in the (66), in the (67) and in the (68) the gravitational mass is erroneously considered to be equal to the sum only of the rest mass and of the contribution to the mass by the kinetic energy, because the contribution of the gravitational potential energy to the determination of the gravitational mass is erroneously considered to be null.

Moreover, according to the erroneous formula (68), if r is small enough in such a way that we have: $\frac{GM}{rc^2} \geq 1$, then there is not any value of velocity $v < c$ for the particle (or small mass) for having a stable circular orbital motion with a radius r; and, obviously, we know that a particle (or small mass) with rest mass greater than zero always has a velocity less than c.

Obviously, the correct formula (65) and the incorrect formula (68) are practically equal to each other for the values of M and r for which we have: $\frac{GM}{rc^2} \ll 1$, that is when we can neglect $\frac{GM}{rc^2}$ with respect to 1. Analogously, we have the same for the correct formula (59) and the incorrect formula (66), that is also these two formulas are practically equal to each other when we can neglect $\frac{GM}{rc^2}$ with respect to 1.

In the case of a photon, instead of $m_0 c^2 \gamma$ we have hf and, moreover, the velocity is always c. For which we have that the analogues of the (60) and of the (49) are respectively:

$$\frac{hf}{c^2}\frac{c^2}{r} = hf\,\frac{1}{r} \qquad (69)$$

$$m_g = \frac{E_T}{c^2} = \frac{hf}{c^2 \gamma_r} = \frac{hf}{c^2} \frac{1}{1 + \dfrac{GM}{rc^2}} \tag{70}$$

On the other hand, the (59) becomes:

$$F_g = \frac{GM}{r^2} \frac{hf}{c^2} \frac{1}{1 + \dfrac{GM}{rc^2}} \tag{71}$$

Now, from the (69) and the (71), if there were a stable circular orbital motion for a photon we would have:

$$\frac{GM}{r^2} \frac{hf}{c^2} \frac{1}{1 + \dfrac{GM}{rc^2}} = \frac{hf}{c^2} \frac{c^2}{r} \tag{72}$$

From which we would have:

$$\frac{GM}{rc^2} \frac{1}{1 + \dfrac{GM}{rc^2}} = 1 \tag{73}$$

Now, we have that M and r are always positive, for which $\left(1 + \dfrac{GM}{rc^2}\right)$ is always greater than zero. Therefore, we can multiply both sides of the equation (73) by $\left(1 + \dfrac{GM}{rc^2}\right)$ and so we would have:

$$\frac{GM}{rc^2} = 1 + \frac{GM}{rc^2} \tag{74}$$

The (74) is obviously equivalent to the equation $0 = 1$, for which the (74) is never satisfied and so also the (73) and consequently also the (72): this implies that there is not any stable circular orbital motion for a photon. All this, obviously, is in accordance with what we have seen, that is with the fact that a photon is always free to go away from a gravitational field due to a spherically symmetric central static mass and therefore a photon cannot have in any case a stable circular orbital motion for a gravitational field due to a spherically symmetric central static mass.

On the other hand, in the case of a photon, the erroneous formulas (66), (67) and (68) become respectively:

$$F_g = \frac{GM}{r^2} \frac{hf}{c^2} \tag{75}$$

$$\frac{GM}{r^2} \frac{hf}{c^2} = \frac{hf}{r} \tag{76}$$

$$\frac{GM}{rc^2} = 1 \tag{77}$$

Obviously, also here, we can note that in the (75), in the (76) and in the (77) the gravitational mass of the photon is erroneously considered to be equal to hf/c^2, that is to the analogue of the sum only of the rest mass and of the contribution to the mass by the kinetic energy in the case of a particle with rest mass greater than zero, because, also here, the contribution of the gravitational potential energy to the determination of the gravitational mass of the photon is erroneously considered to be null.

Moreover, according to the erroneous formula (77) a photon would have a stable circular orbital motion at a value of r equal to $\dfrac{GM}{c^2}$.

Obviously, when the values of M and r are such that we have: $\dfrac{GM}{rc^2} \ll 1$, that is when we can neglect $\dfrac{GM}{rc^2}$ with respect to 1, the correct formula (73) and the incorrect formula (77) are practically equal to each other and are in agreement to each other as for the fact that they are without any solution. Analogously, also the correct formula (71) and the incorrect formula (75) are practically equal to each other when we can neglect $\dfrac{GM}{rc^2}$ with respect to 1.

CHAPTER III
THE DYNAMICS IN THE FIELD OF APPLICATION OF THE GENERAL THEORY OF RELATIVITY

1 The gravitational field due to a spherically symmetric static mass according to the General Theory of Relativity

1.1 The usual Schwarzschild solution

For a gravitational field due to a spherically symmetric static central mass M, the currently used solution of the Einstein's field equation (of the General Theory of Relativity)[2] is the so-called Schwarzschild solution[3]:

$$ds^2 = \left(1 - \frac{2GM}{rc^2}\right) c^2 dt^2 - \frac{1}{1 - \frac{2GM}{rc^2}}\, dr^2 +$$

---

[2] Cf. EINSTEIN (1916); OHANIAN – RUFFINI pp. 275-293; MISNER – THORNE – WHEELER pp. 383-590; WEINBERG (1972) pp. 151-171; ANDERSON pp. 329-371.

[3] Cf. SCHWARZSCHILD; OHANIAN – RUFFINI pp. 293-298; MISNER – THORNE – WHEELER pp. 636-687.819-840; WEINBERG (1972) pp. 175-182; ANDERSON pp. 381-392.

$$- r^2 d\theta^2 - r^2 \sin^2 \theta \, d\varphi^2 \qquad\qquad (78)$$

This metric would imply the presence of an event horizon at the so-called Schwarzschild radius $r_s = \dfrac{2GM}{c^2}$ and of a black hole inside that event horizon, but this is clearly against what we have seen in our general treatment. In fact, we have seen that, in general, there is not any event horizon and consequently there is not any black hole in the case of a gravitational field due to a spherically symmetric static central mass. Furthermore, we have seen that the presence of an event horizon would imply the violation of the principle of conservation of energy, since a particle that goes inside an event horizon has not in any case the energy sufficient for going away from the inside of this same event horizon.

Moreover, the Einstein's field equation of the General Theory of Relativity[4] is symmetric with respect to time[5]. In other words, we cannot use the General Theory of Relativity for justifying the existence of the arrow of time, because the symmetry of the equations of the General Theory of Relativity implies that all the solutions of these equations are symmetric with respect to time and reversible over time. In particular, the motion back in time (of a particle) is always equivalent to a motion allowed by the equations of the General Theory of Relativity[6].

Therefore, any solution of the Einstein's field equation (of the General Theory of Relativity) cannot include an event

---

[4] Cf. EINSTEIN (1916); OHANIAN – RUFFINI pp. 275-293; MISNER – THORNE – WHEELER pp. 383-590; WEINBERG (1972) pp. 151-171; ANDERSON pp. 329-371.

[5] Cf. HAWKING (1997) pp. 97.167.176; DAVIES (1997) pp. 229-231; BARROW (2003) pp. 284-286; BARROW (1996) pp. 167-168; PENROSE (2009) pp. 386-389; GOTT p. 197.

[6] Cf. PACE (2016b) pp. 79-80.

horizon, since an event horizon is evidently not symmetric with respect to time, and consequently cannot include a black hole. In other words, the symmetry with respect to time of the Einstein's field equation (of the General Theory of Relativity) implies the impossibility of the existence of any event horizon due to gravitational fields and consequently the impossibility of the existence of any black hole due to gravitational fields[7].

Moreover, the (78) would imply the presence of a strange singularity at the Schwarzschild radius $r_S = \dfrac{2GM}{c^2}$: in fact, the greater M is, the smaller the gravitational force at the Schwarzschild radius is; and for $M \to + \infty$ we have that the gravitational force at the Schwarzschild radius tends to zero, but, according to the (78), this singularity would be always present (at the Schwarzschild radius).

Furthermore, according to the (78) for values of r less than the Schwarzschild radius $r_S = \dfrac{2GM}{c^2}$ the coefficient of $c^2 dt^2$ would become negative and the coefficient of dr^2 would become positive.

All this implies that the (78) cannot be a true solution of the Einstein's field equation of the General Theory of Relativity.

On the other hand, the correct solutions of the Einstein's field equation of the General Theory of Relativity that I have recently found[8] (instead of the currently used Schwarzschild, Kerr, Reissner-Nordstrøm and Kerr-Newman erroneous solutions both in the case with $\Lambda = 0$ and in the case with $\Lambda > 0$, where Λ is the cosmological constant) confirm the fact that, according to the General Theory of Relativity, there is not any event horizon and consequently there is not any black hole[9]. For the

---

[7] Cf. PACE (2016b) pp.79-80.
[8] Cf. PACE (2016b) pp. 7-78.
[9] Cf. PACE (2016b) pp. 7-78.

convenience of the reader I will report also here my derivation of the correct Schwarzschild solution[10].

1.2 The correct Schwarzschild solution

For calculating the correct Schwarzschild solution (of the Einstein's field equation of the General Theory of Relativity) I will correct the treatment that has been made by H. C. Ohanian and R. Ruffini for obtaining the Schwarzschild solution (of the Einstein's field equation of the General Theory of Relativity)[11].

The Schwarzschild solution (of the Einstein's field equation of the General Theory of Relativity) is the solution (of the Einstein's field equation of the General Theory of Relativity) for the gravitational field surrounding a spherically symmetric mass distribution.

Therefore, for obtaining the correct Schwarzschild solution, in what follows we will solve the Einstein's equations, or more precisely the Einstein's field equation, of the General Theory of Relativity in the exterior of a spherically symmetric mass distribution, that is, in the vacuum region surrounding the spherically symmetric mass distribution. For simplicity in this case we will use the formulas with $c \equiv 1$.

Now, if the mass distribution is spherically symmetric, the gravitational field generated by this mass distribution must also have spherical symmetry. Since it is sometimes of interest to deal with a collapsing spherical mass – that is, a sphere that shrinks in size – we will begin with an *Ansatz* (that is, with an approach) for the space-time metric that includes a time dependence for the components of the metric tensor. However, at the end of our calculation we will find that in the exterior region this time

---

[10] Cf. Pace (2016b) pp. 7-25.
[11] Cf. Ohanian – Ruffini pp. 293-298.

dependence disappears, because, as in the case of the Newtonian gravitational potential, the exterior solution for the metric tensor of a spherical mass does not depend on the size of the mass and remains static even when the mass collapses.

The spherical symmetry of the gravitational field imposes severe restrictions on the form of the space-time interval. The rectangular displacements dx, dy, dz must occur in the combination $dx^2 + dy^2 + dz^2$ characteristic of spherical symmetry; this combination takes the familiar form $dr^2 + r^2 d\theta^2 + r^2 \sin^2 \theta \, d\varphi^2$ when is expressed in spherical polar coordinates. Furthermore, although the product $drdt$ is consistent with spherical symmetry, the products $d\theta dt$ and $d\varphi dt$ are not consistent with spherical symmetry, because they would imply that a light signal (with $ds^2 = 0$) has different speeds when moving in the direction of increasing or decreasing θ and φ. Accordingly, the space-time interval must be of the form:

$$ds^2 = A(r,t)dt^2 +$$

$$- B(r,t)[dr^2 + r^2 d\theta^2 + r^2 \sin^2 \theta \, d\varphi^2] +$$

$$- 2F(r,t)drdt \tag{79}$$

where $A(r,t)$, $B(r,t)$ and $F(r,t)$ are some functions of the radial coordinate r and the time coordinate t.

The angular coordinates θ and φ are unambiguous; the measurement of these coordinates depends only on our ability to divide a circumference concentric with the mass into equal parts, which we can do even when the functions $A(r,t)$, $B(r,t)$ and $F(r,t)$ are not specified. The radial coordinate r is ambiguous because we do not yet know its precise relation to the measurement of distance. For a start, we will treat r simply as a parameter that identifies different spherical surfaces concentric with the mass. The time coordinate t suffers from similar ambiguities. However, we will insist that when $r \to +\infty$ and the space-time becomes flat, the increments in r and t should equal the increments in the true distance and time, respectively. This requires that for $r \to +\infty$, $A(r,t)$ and $B(r,t) \to 1$.

The term involving $drdt$ in the equation (79) can be eliminated by a change in the time coordinate. We introduce a new time coordinate $\tilde{t}$ such that:

$$d\tilde{t} = (Adt - Fdr)Q(r,t) \tag{80}$$

where $Q(r,t)$ is a function of r and t that is to be chosen so as to make the right side of the equation (80) into a perfect differential [this requires that Q satisfy the following differential equation: $(\partial/\partial t)(FQ) = -(\partial/\partial r)(AQ)$].

From the equation (80) we obtain:

$$Adt^2 - 2Fdrdt = \frac{1}{Q^2 A}d\tilde{t}^2 - \frac{F^2}{A}dr^2 \tag{81}$$

which gives:

$$ds^2 = \frac{1}{Q^2 A}d\tilde{t}^2 - \left(B + \frac{F^2}{A}\right)dr^2 +$$

$$- B[r^2 d\theta^2 + r^2 \sin^2\theta \, d\varphi^2] \tag{82}$$

We can simplify the equation (82) further by introducing a new radial coordinate $\tilde{r} = r\sqrt{B(r)}$. This has the advantage that the terms involving angular displacements reduce to $\tilde{r}^2 d\theta^2 + \tilde{r}^2 \sin^2\theta\, d\varphi^2$, so we obtain:

$$ds^2 = \frac{1}{Q^2 A} d\tilde{t}^2 - \left(B + \frac{F^2}{A}\right)\left(\frac{\partial r}{\partial \tilde{r}}\right)^2 d\tilde{r}^2 +$$

$$- \tilde{r}^2 d\theta^2 - \tilde{r}^2 \sin^2\theta\, d\varphi^2 \tag{83}$$

For the solution of the Einstein's equations, or more precisely of the Einstein's field equation, of the General Theory of Relativity, it is convenient to omit the tildes in the equation (83) and to write the unknown functions multiplying dt^2 and dr^2 as exponentials:

$$ds^2 = e^{N(r,t)} dt^2 - e^{L(r,t)} dr^2 - r^2 d\theta^2 +$$

$$- r^2 \sin^2\theta\, d\varphi^2 \tag{84}$$

With $x^0 = t$, $x^1 = r$, $x^2 = \theta$, $x^3 = \varphi$, the metric tensor corresponding to the equation (84) is:

$$g_{\mu\nu} =$$

$$
= \begin{pmatrix} e^N & 0 & 0 & 0 \\ 0 & -\,e^L & 0 & 0 \\ 0 & 0 & -\,r^2 & 0 \\ 0 & 0 & 0 & -\,r^2 \sin^2\theta \end{pmatrix}
$$

$$(85)$$

The unknown functions are now $N(r,t)$ and $L(r,t)$. We will use the Einstein's equations, or more precisely the Einstein's field equation, of the General Theory of Relativity to find them.

The Christoffel symbols for a metric tensor of the form (85) are (the primes indicate derivatives, $N' \equiv \frac{\partial N}{\partial r}$, $L' \equiv \frac{\partial L}{\partial r}$; the dots indicate time derivatives, $\dot{N} \equiv \frac{\partial N}{\partial t}$, $\dot{L} \equiv \frac{\partial L}{\partial t}$):

$$\Gamma^0{}_{00} = \frac{1}{2}\,\dot{N} \tag{86}$$

$$\Gamma^0{}_{01} = \Gamma^0{}_{10} = \frac{1}{2}\,N' \tag{87}$$

$$\Gamma^0{}_{11} = \frac{1}{2}\,\dot{L}\,e^{L-N} \tag{88}$$

$$\Gamma^1{}_{00} = \frac{1}{2}\,N'e^{N-L} \tag{89}$$

$$\Gamma^1{}_{10} = \Gamma^1{}_{01} = \frac{1}{2}\dot{L} \tag{90}$$

$$\Gamma^1{}_{11} = \frac{1}{2}L' \tag{91}$$

$$\Gamma^1{}_{22} = -r\,e^{-L} \tag{92}$$

$$\Gamma^1{}_{33} = -r\sin^2\theta\,e^{-L} \tag{93}$$

$$\Gamma^2{}_{12} = \Gamma^2{}_{21} = \frac{1}{r} \tag{94}$$

$$\Gamma^2{}_{33} = -\sin\theta\cos\theta \tag{95}$$

$$\Gamma^3{}_{13} = \Gamma^3{}_{31} = \frac{1}{r} \tag{96}$$

$$\Gamma^3{}_{23} = \Gamma^3{}_{32} = \cot\theta \tag{97}$$

All other Christoffel symbols are zero. The Ricci tensor $R_{\mu\nu}$ can be calculated from the Christoffel symbols; only the 00, 01, 10, 11, 22, 33 components of $R_{\mu\nu}$ are nonzero.

The Einstein's field equation (of the General Theory of Relativity) in vacuum is:

$$R_{\mu\nu} - \frac{1}{2} g_{\mu\nu} R = 0 \tag{98}$$

By taking the trace of this equation, we find that $R = 0$. Hence the equation (98) reduces to:

$$R_{\mu\nu} = 0 \tag{99}$$

The 00, 01, 10, 11, 22 and 33 components of this equation are as follows:

$$R_{00} = e^{N-L}\left(-\frac{N''}{2} + \frac{L'N'}{4} - \frac{N'^2}{4} - \frac{N'}{r}\right) +$$

$$+ \frac{\ddot{L}}{2} + \frac{\dot{L}(\dot{L} - \dot{N})}{4} = 0 \tag{100}$$

$$R_{01} = -R_{10} = -\frac{\dot{L}}{r} = 0 \tag{101}$$

$$R_{11} = \frac{N''}{2} - \frac{L'N'}{4} + \frac{N'^2}{4} - \frac{L'}{r} +$$

$$+ e^{L-N}\left[\frac{\ddot{L}}{2} + \frac{\dot{L}(\dot{L} - \dot{N})}{4}\right] = 0 \tag{102}$$

$$R_{22} = e^{-L} \left[1 + \frac{1}{2} r (N' - L') \right] +$$

$$- 1 = 0 \tag{103}$$

$$R_{33} = \sin^2 \theta \, e^{-L} \left[1 + \frac{1}{2} r (N' - L') \right] +$$

$$- \sin^2 \theta = 0 \tag{104}$$

Although these equations look complicated, they can be integrated quite easily. The equation (101) says that L is time independent, and this implies that *all* the terms involving time derivatives in the equations (100) and (102) drop out. The sum of the equation (102) and e^{L-N} times the equation (100) then gives:

$$- \frac{L'}{r} - \frac{N'}{r} = 0 \tag{105}$$

from which we have:

$$N = - L + h(t) \tag{106}$$

where $h(t)$ is an arbitrary function of time [in regard to the space derivatives in the equation (105), this function $h(t)$ behaves like a constant of integration].

When we substitute the equation (106) into the equation (103), we find:

$$e^{-L}\left(-rL' + 1\right) = 1 \tag{107}$$

Since L is independent from time, the partial derivative of L with respect to r is equal to a simple derivative of L with respect to r. Therefore the equation (107) is equal to:

$$e^{-L}\left(-r\frac{dL}{dr} + 1\right) = 1 \tag{108}$$

From which we have:

$$-r\,e^{-L}\frac{dL}{dr} + e^{-L} = 1 \tag{109}$$

from which we have:

$$-r\,e^{-L}\frac{dL}{dr} = 1 - e^{-L} \tag{110}$$

from which we have:

$$-e^{-L}\frac{dL}{1 - e^{-L}} = \frac{dr}{r} \tag{111}$$

from which we have:

$$-e^{-L}\frac{dL}{e^{-L}(e^{L} - 1)} = \frac{dr}{r} \tag{112}$$

from which, finally, we have:

$$\frac{dL}{1 - e^L} = \frac{dr}{r} \qquad (113)$$

By integrating the two parts of the equation (113), we have:

$$\int \frac{dL}{1 - e^L} = \int \frac{dr}{r} \qquad (114)$$

From which we have:

$$L - \ln(|1 - e^L|) = \ln r + K \qquad (115)$$

where K is a constant and in particular a real number since we are operating in the field of the real numbers and not in the field of the imaginary numbers or (in the field) of the complex numbers. In fact we can note that in the equation (114) e^L is an element changed of sign of the metric tensor that is expressed by the equation (96), for which e^L must be always a real number. On the other hand, r must be always a real number because represents the radius of polar coordinates. Analogously dr must be always a real number because it (dr) is the infinitesimal variation of a real number. Moreover, from the equation (113), because of the fact that e^L, r and dr are all real, we can say that also dL must be real. For which when we integrate the equation (114) the integration is done in the real field both in dL and in dr: the integration constant must, therefore, be a real number. As we will see, the fact that K is a real number is very important.

Therefore, from the equation (115) we have:

$$L - K = \ln r + \ln(|1 - e^L|) \tag{116}$$

From which we have two possibilities: $1 - e^L > 0$ and $1 - e^L < 0$.

In the case of $1 - e^L > 0$ we have:

$$L - K = \ln\{r\,(1 - e^L)\} \tag{117}$$

from which we have:

$$e^{-K}\,e^L = r\,(1 - e^L) \tag{118}$$

from which we have:

$$e^L(e^{-K} + r) = r \tag{119}$$

from which we have:

$$e^L = \frac{r}{e^{-K} + r} \tag{120}$$

from which, finally, we have:

$$e^L = \frac{1}{1 + \dfrac{e^{-K}}{r}} \tag{121}$$

where e^{-K} is a positive real number, since e^{x} is positive for any real number x, and $-K$ is a real number. Moreover, we can see that according to the equation (121) e^{L} is always < 1 for any value of $r > 0$, for which $1 - e^{L}$ is always > 0 for any value of $r > 0$. [In fact, we must remember that we have obtained this formula (121) in the hypothesis of $1 - e^{L} > 0$].

Instead in the case of $1 - e^{L} < 0$ we have:

$$L - K = \ln\{r\,(e^{L} - 1)\} \tag{122}$$

from which we have:

$$e^{-K}\,e^{L} = r\,(e^{L} - 1) \tag{123}$$

from which we have:

$$e^{L}(e^{-K} - r) = -r \tag{124}$$

from which we have:

$$e^{L} = \frac{-r}{e^{-K} - r} \tag{125}$$

from which, finally, we have:

$$e^{L} = \frac{1}{1 - \dfrac{e^{-K}}{r}} \tag{126}$$

Now this formula (126) is valid, obviously, only in the case of $1 - e^L < 0$ [in fact, we must remember that we have obtained this formula (126) in the hypothesis of $1 - e^L < 0$], that is in the case of $e^L > 1$, and therefore the formula (126) cannot be valid for values of $r < e^{-K}$ because for these values (of r) e^L is < 0 according to the formula (126). On the other hand, we already know that e^{-K} is a positive real number, since e^x is positive for any real number x and $-K$ is a real number.

Since we search a solution that can be valid for all the values of $r > 0$ we must reject the solution (126), because the solution (126) is not valid for all the values of $r > 0$. Therefore, we have that the only acceptable solution is the formula (121). On the other hand, we will see also another reason for discarding the equation (126).

Now we can simplify the equation (121) by means of introducing a positive real constant C, that is equal by definition to e^{-K}:

$$C \equiv e^{-K} \tag{127}$$

Therefore the equation (121) becomes:

$$e^L = \frac{1}{1 + \dfrac{C}{r}} \tag{128}$$

where C is (a constant that is equal to) a positive real number, in particular C is > 0.

According to the equation (106), the solution for $N(r)$ is then:

$$e^{N(r)} = e^{-L(r)+h(t)} = \left(1 + \frac{C}{r}\right)e^{h(t)} \qquad (129)$$

With the solutions (128) and (129), the space-time interval for the spherically symmetric field takes the form:

$$ds^2 = \left(1 + \frac{C}{r}\right)e^{h(t)}dt^2 - \frac{1}{1 + \dfrac{C}{r}}\,dr^2 +$$

$$- r^2 d\theta^2 - r^2 \sin^2\theta\, d\varphi^2 \qquad (130)$$

where C is (a constant that is equal to) a positive real number, in particular C is > 0.

Here, the time-dependent function $h(t)$ remains undetermined and unknown. The field equations do not determine this function, but we can eliminate it by means of a transformation of the time coordinate. If we adopt a new time coordinate $\tilde{t}$ such that $d\tilde{t} \equiv e^{\frac{h(t)}{2}}dt$, then the function $h(t)$ disappears from view, and the first term of the right side of the equation (130) becomes $\left(1 + \frac{C}{r}\right)d\tilde{t}^2$. We can then omit the tilde, so we obtain:

$$ds^2 = \left(1 + \frac{C}{r}\right)dt^2 - \frac{1}{1 + \dfrac{C}{r}}\,dr^2 +$$

$$- r^2 d\theta^2 - r^2 \sin^2\theta\, d\varphi^2 \qquad (131)$$

Here, we underline again that C is (a constant that is equal to) a positive real number, in particular C is > 0. Commonly instead of C the authors write[12] $- C$ and then consider $- C$ equal to a negative number, in particular to $- 2GM$: now, this is erroneous, as we have seen. In other words, the formula commonly used instead of the expression (131) is not a true solution of the Einstein's field equation of the General Theory of Relativity, if that formula is interpreted in the usual way.

To find the value of C in the equation (131), we must compare this expression (131) with the result obtained from the linear theory. Now, the formula (obtained from the linear theory) commonly used for the comparison is[13]:

$$ds^2 = \left(1 - \frac{2GM}{r}\right) dt^2 - \left(1 + \frac{2GM}{r}\right) (dx^2 +$$

$$+ dy^2 + dz^2) \tag{132}$$

But this formula (obtained from the linear theory) is not expressed as a function of the coordinates measured in the presence of the gravitational fields (that is, measured relatively to a reference frame that is integral with a space-time curved for the presence of the gravitational fields) but as a function of the coordinates measured in the absence of any gravitational field (that is, measured relatively to a reference frame that is integral with a flat space-time)[14], while the expression (131) is expressed as a function of the coordinates measured in the presence of the gravitational fields (that is, measured relatively to a reference frame that is integral with a space-time curved for the presence of

---

[12] Cf. OHANIAN – RUFFINI pp. 293-298.
[13] Cf. OHANIAN – RUFFINI pp. 293-298.
[14] Cf. OHANIAN – RUFFINI pp. 130-131.

the gravitational fields), since the Einstein's field equation of the General Theory of Relativity is expressed as a function of the coordinates measured in the presence of the gravitational fields (that is, measured relatively to a reference frame that is integral with a space-time curved for the presence of the gravitational fields) and we have obtained the expression (131) by starting from the Einstein's field equation of the General Theory of Relativity. Therefore we, instead of the expression (132), need to have the expression (obtained from the linear theory) of the space-time interval, which (space-time interval) be expressed as a function of the coordinates measured in the presence of the gravitational fields (that is, measured relatively to a reference frame that is integral with a space-time curved for the presence of the gravitational fields). In other words, the expression (132) represents the metric that is formally flat when is expressed as a function of the coordinates measured in the presence of the gravitational fields (that is, measured relatively to a reference frame that is integral with a space-time curved for the presence of the gravitational fields) which metric is expressed as a function of the coordinates measured in the absence of any gravitational field (that is, measured relatively to a reference frame that is integral with a flat space-time). Instead we need to have the formula (obtained from the linear theory) which (formula) represent the metric that is formally flat when is expressed as a function of the coordinates measured in the absence of any gravitational field (that is, measured relatively to a reference frame that is integral with a flat space-time), which metric be expressed as a function of the coordinates measured in the presence of the gravitational fields (that is, measured relatively to a reference frame that is integral with a space-time curved for the presence of the gravitational fields).

On the other hand, from the formula (132) we have:

$$d\tau^2 = \left(1 - \frac{2GM}{r}\right) dt^2 \qquad (133)$$

where $d\tau$ is the time measured by clocks positioned in the gravitational field (that is, measured relatively to a reference frame that is integral with a space-time curved for the presence of the gravitational field). And therefore for $2GM \ll r$ we have:

$$dt^2 \cong \left(1 + \frac{2GM}{r}\right) d\tau^2 \qquad (134)$$

And analogously, from the formula (132) we have relatively to the spatial coordinates:

$$\left(1 + \frac{2GM}{r}\right)(dx^2 + dy^2 + dz^2) =$$

$$= \{dx_g^2 + dy_g^2 + dz_g^2\} =$$

$$= \{dr^2 + r^2 d\theta^2 + r^2 \sin^2 \theta \, d\varphi^2\} \qquad (135)$$

And therefore for $2GM \ll r$ we have:

$$(dx^2 + dy^2 + dz^2) =$$

$$\cong \left(1 - \frac{2GM}{r}\right)\{dx_g^2 + dy_g^2 + dz_g^2\} =$$

$$= \left(1 - \frac{2GM}{r}\right) \{dr^2 + r^2 d\theta^2 +$$

$$+ r^2 \sin^2 \theta \, d\varphi^2\} \tag{136}$$

where x_g, y_g, z_g or r, ϑ, φ are the spatial coordinates measured by instrumentation (that is) positioned in the gravitational field (that is, measured relatively to a reference frame that is integral with a space-time curved for the presence of the gravitational field).

Therefore the expression (131) in the limit for $r \to \infty$ must be compared not with the expression (132) but with this expression:

$$ds^2 = \left(1 + \frac{2GM}{r}\right) dt^2 +$$

$$- \left(1 - \frac{2GM}{r}\right) (dx^2 + dy^2 + dz^2) \tag{137}$$

or better with this other expression:

$$ds^2 = \left(1 + \frac{2GM}{r}\right) dt^2 - \left(1 - \frac{2GM}{r}\right) (dr^2 +$$

$$+ r^2 d\theta^2 + r^2 \sin^2 \theta \, d\varphi^2) \tag{138}$$

In fact the formulas (137) and (138), in the case of $\frac{2GM}{r} \ll$ 1, represent the expression (obtained from the linear theory) of the space-time interval, which (space-time interval) is expressed as a function of the coordinates measured in the presence of the gravitational fields (that is, measured relatively to a reference frame that is integral with a space-time curved for the presence of the gravitational fields), or in other words represent the expression (obtained from the linear theory) of the metric that is formally flat when is expressed as a function of the coordinates measured in the absence of any gravitational field (that is, measured relatively to a reference frame that is integral with a flat space-time) which metric is expressed as a function of the coordinates measured in the presence of the gravitational fields (that is, measured relatively to a reference frame that is integral with a space-time curved for the presence of the gravitational fields).

Since we cannot compare the expression (131) with the expression (137) or with the expression (138) directly, because they [the expressions (137) and (138)] have a form which is different from that of the expression (131), let us change the equation (131) by introducing a new coordinate $\tilde{r}$:

$$\tilde{r} \equiv \frac{1}{2}\sqrt{r^2 + Cr} + \frac{1}{2}r + \frac{1}{4}C \tag{139}$$

or, equivalently,

$$r \equiv \tilde{r}\left(1 - \frac{C}{4\tilde{r}}\right)^2 \tag{140}$$

This transformation gives:

$$ds^2 = \left(\frac{1 + \frac{C}{4\tilde{r}}}{1 - \frac{C}{4\tilde{r}}}\right)^2 dt^2 - \left(1 - \frac{C}{4\tilde{r}}\right)^4 \{d\tilde{r}^2 + \\ + \tilde{r}^2 d\theta^2 + \\ \tilde{r}^2 \sin^2\theta\, d\varphi^2\} \tag{141}$$

The coordinates used in the expression (141) are called *isotropic*.

In the weak field limit ($\tilde{r} \to \infty$), the equation (141) reduces to:

$$ds^2 = \left(1 + \frac{C}{\tilde{r}}\right) dt^2 - \left(1 - \frac{C}{\tilde{r}}\right)\{d\tilde{r}^2 + \\ + \tilde{r}^2 d\theta^2 + \tilde{r}^2 \sin^2\theta\, d\varphi^2\} \tag{142}$$

Obviously, this equation has the same form of the equation (138) [or of the equation (137)]. By comparing the equation (142) with the equation (138), we see that:

$$C = 2GM \tag{143}$$

We can note that this effectively is in accord with the limitation that C must be a positive real number. (The case with $M \to 0$ corresponds to $K \to +\infty$ and $C \equiv e^{-K} \to 0$). On the other hand, this obviously is another reason for discarding the solution (126).

By means of the substitution of C with $2GM$, according to (143), the equation (131) becomes:

$$ds^2 = \left(1 + \frac{2GM}{r}\right)dt^2 - \frac{1}{1 + \dfrac{2GM}{r}}\,dr^2 +$$

$$- r^2 d\theta^2 - r^2 \sin^2\theta\, d\varphi^2 \qquad (144)$$

where M is (a constant, that is equal to) the total mass of the system[15].

For simplicity we have used the formulas with $c \equiv 1$ for calculating the correct Schwarzschild solution. Obviously, in the case more general (in which c is not defined equal to 1) we have that the (144) becomes:

$$ds^2 = \left(1 + \frac{2GM}{rc^2}\right)c^2 dt^2 - \frac{1}{1 + \dfrac{2GM}{rc^2}}\,dr^2 +$$

$$- r^2 d\theta^2 - r^2 \sin^2\theta\, d\varphi^2 \qquad (145)$$

This is the correct final form of the Schwarzschild solution, that is the correct solution for the case which Schwarzschild has analysed.

Instead, as we have seen, the common erroneous expression for the Schwarzschild solution is[16]:

---

[15] Cf. PACE (2016b) p. 24.
[16] Cf. OHANIAN – RUFFINI p. 298.

$$ds^2 = \left(1 - \frac{2GM}{rc^2}\right)c^2 dt^2 - \frac{1}{1 - \dfrac{2GM}{rc^2}}\, dr^2 +$$

$$- r^2 d\theta^2 - r^2 \sin^2 \theta \, d\varphi^2 \qquad (146)$$

The difference between the correct formula (145) and the incorrect formula (146) substantially consists in two plus signs instead of two minus signs. As we will see, this entails enormous physical consequences.

1.3 The consequences of the correct Schwarzschild solution

Now, we have seen that the correct Schwarzschild solution of the Einstein's field equation (of the General Theory of Relativity) is this other expression:

$$ds^2 = \left(1 + \frac{2GM}{rc^2}\right)c^2 dt^2 - \frac{1}{1 + \dfrac{2GM}{rc^2}}\, dr^2 +$$

$$- r^2 d\theta^2 - r^2 \sin^2 \theta \, d\varphi^2 \qquad (147)$$

We can easily see that this metric does not imply any event horizon and, consequently, this metric does not imply any black hole[17].

---

[17] Cf. OHANIAN – RUFFINI pp. 324-342; MISNER – THORNE – WHEELER pp. 842-940; WEINBERG (1972) pp. 207-208.297.349.479.

In fact, according to the (147), there is not any singularity at the Schwarzschild radius $r_s = \dfrac{2GM}{c^2}$ and in general at any value of $r > 0$ [18].

Moreover, according to the (147), for any value of $r > 0$ (and in particular for any value of r less than the Schwarzschild radius $r_s = \dfrac{2GM}{c^2}$) the coefficient of $c^2 dt^2$ always remains positive and the coefficient of dr^2 always remains negative. Therefore, since the coefficient of $c^2 dt^2$ is always not negative the time (that is, the temporal coordinate) does not become in any case as a spatial coordinate, contrary to the common treatment of the space-time inside the event horizon [19]. On the other hand, since the coefficient of dr^2 is always not positive the spatial coordinate r does not become in any case as a temporal coordinate, contrary to the common treatment of the space-time inside the event horizon [20].

Consequently, the light cones are always orientated in the usual way, in particular there is not any horizontal inclination of the light cones, contrary to the common treatment of the space-time inside the event horizon [21].

On the other hand, as I have already shown in another publication [22], the fact that the coefficient of $c^2 dt^2$ in the equation (147) is > 1 in the presence of a gravitational field entails that in the presence of a gravitational field the clocks go more slowly (that is, that the measurements of time relatively to a reference frame that is integral with a space-time curved for the

---

[18] Cf. PACE (2016b) pp. 36-39.

[19] Cf. OHANIAN – RUFFINI pp. 324-342; MISNER – THORNE – WHEELER pp. 819-940; PACE (2016b) p. 38.

[20] Cf. OHANIAN – RUFFINI pp. 324-342; MISNER – THORNE – WHEELER pp. 819-940; PACE (2016b) p. 38.

[21] Cf. OHANIAN – RUFFINI pp. 324-342; MISNER – THORNE – WHEELER pp. 819-940; PACE (2016b) pp. 38-39.

[22] Cf. PACE (2016b) pp. 25-29.

presence of the gravitational field flow more slowly): and we know that this is in accord with the experimental results. In fact, we have:

$$\left(1 + \frac{2GM}{rc^2}\right) d\tau^2 = dt^2 \qquad (148)$$

For which the time τ measured by a clock positioned in the gravitational field (that is, measured relatively to a reference frame that is integral with a space-time curved for the presence of the gravitational field) is less than the time t measured by a clock positioned where there is not any gravitational field (that is, measured relatively to a reference frame that is integral with a flat space-time).

Analogously, the fact that the coefficient of $c^2 dt^2$ in the equation (147) has the form $\left(1 + \frac{2GM}{rc^2}\right)$ entails that the clocks positioned in a more intense gravitational field go more slowly than the clocks positioned in a less intense gravitational field (that is, that the measurements of time relatively to a reference frame that is integral with a space-time curved for the presence of a more intense gravitational field flow more slowly than the measurements of time relatively to a reference frame that is integral with a space-time curved for the presence of a less intense gravitational field). In fact we have:

$$\left(1 + \frac{2GM}{r_1 c^2}\right) d\tau_1{}^2 = \left(1 + \frac{2GM}{r_2 c^2}\right) d\tau_2{}^2 \qquad (149)$$

Now, when the gravitational field is more intense (that is, when r is more small) the coefficient of $d\tau^2$ is greater, for which the corresponding time flows more slowly (that is, the measurements of time relatively to a reference frame that is integral with a space-time curved for the presence of a more intense gravitational field flow more slowly than the measurements of time relatively to a reference frame that is integral with a space-time curved for the presence of a less intense gravitational field). In particular, we have:

$$\frac{d\tau_2}{d\tau_1} = \sqrt{\frac{1 + \dfrac{2GM}{r_1 c^2}}{1 + \dfrac{2GM}{r_2 c^2}}} \qquad (150)$$

For which when $r_2 > r_1$, that is when the second gravitational field is less intense that the first gravitational field, then $d\tau_2 > d\tau_1$: we have the well-known result (that has been confirmed by numerous experiments) that the more intense is the gravitational field the more slowly the clocks go (that is, that the measurements of time relatively to a reference frame that is integral with a space-time curved for the presence of a more intense gravitational field flow more slowly than the measurements of time relatively to a reference frame that is integral with a space-time curved for the presence of a less intense gravitational field).

Analogously, the ratio of the relative frequencies $v_1 = 1/d\tau_1$ and $v_2 = 1/d\tau_2$ is:

$$\frac{\nu_1}{\nu_2} = \sqrt{\frac{1 + \dfrac{2GM}{r_1 c^2}}{1 + \dfrac{2GM}{r_2 c^2}}} \tag{151}$$

This is the correct formula for the gravitational redshift of light in the correct Schwarzschild geometry. We can note that the two formulas (150) and (151) are different from those obtained usually from the incorrect expression of the Schwarzschild solution by means of an incorrect procedure[23], but are equal at the first order in $\dfrac{2GM}{rc^2}$ (to the incorrect formulas). In fact, the formulas commonly used, instead of the two correct formulas (150) and (151), are respectively:

$$\frac{d\tau_2}{d\tau_1} = \sqrt{\frac{1 - \dfrac{2GM}{r_2 c^2}}{1 - \dfrac{2GM}{r_1 c^2}}} \tag{152}$$

$$\frac{\nu_1}{\nu_2} = \sqrt{\frac{1 - \dfrac{2GM}{r_2 c^2}}{1 - \dfrac{2GM}{r_1 c^2}}} \tag{153}$$

---

[23] Cf. OHANIAN – RUFFINI pp. 293-298.308-309.

Now, according to the General Theory of Relativity, the metric described by the (147) is equivalent to the presence of a gravitational field. Consequently, we could describe this situation as a gravitational field in a flat metric (of the spacetime), with a gravitational potential dependent only on r, in a similar way to what we have made in the general case.

Therefore, as in the general case, for every r there is an escape velocity v_r dependent only on the variable r and less than the velocity of light in vacuum c; moreover a photon is always free to go away from the gravitational field.

On the other hand, as in the general case, the gravitational mass is given by the (3), and therefore by the (24) for the particles with rest mass greater than zero and by the (35) for the photons. Therefore, for the principle of conservation of energy a particle in motion under the action only of a gravitational field due to a spherically symmetric static central mass always conserves the same total energy and consequently the same gravitational mass.

Also here, as in the general case, we can note that the total energy of a particle in a gravitational field due to a spherically symmetric static central mass is always positive.

2 Extension to any gravitational field due to a static mass distribution (without electric charge)

Now, every gravitational field due to a static mass distribution (without electric charge) is a sum (at the limit an infinite sum) of gravitational fields due to a spherically symmetric static mass, in general with different centres. In fact these gravitational fields (due to a spherically symmetric static mass) can take into account (by means of opportune modifications) also the interaction between the different parts of the mass distribution. Moreover, every gravitational field due to a spherically symmetric static mass does not imply any event horizon and, therefore, does not imply any black hole. Therefore, also a sum (at the limit an infinite sum) of gravitational fields due to a spherically symmetric static mass, in general with different centres, does not imply any event horizon and, therefore, does not imply any black hole. Therefore, any gravitational field due to a static mass distribution (without electric charge) does not imply any event horizon and, therefore, does not imply any black hole.

Moreover, according to the General Theory of Relativity, every gravitational field can be described as a curved metric (of the spacetime), and every curved metric (of the spacetime) is equivalent to the presence of a gravitational field. Consequently, according to the General Theory of Relativity, in general we can describe any gravitational field due to a static mass distribution (without electric charge) in a flat metric (of the spacetime), with a gravitational potential dependent only on the spatial coordinates of every point, in a similar way to what we have made in the case of a gravitational field due to a spherically symmetric central static mass.

Therefore, as in the case due to a spherically symmetric central static mass, also in the case of any gravitational field due to a static mass distribution (without electric charge) for every point there is an escape velocity v_e, but in this case in general v_e is dependent on all the spatial coordinates of the point. However, in any point this v_e is less than the velocity of light in vacuum c. Moreover, also in this case a photon is always free to go away from the gravitational field.

On the other hand, also in this case more general, as in the case due to a spherically symmetric central static mass, the gravitational mass is given by the (3), and therefore by an analogue of the (24) for the particles with rest mass greater than zero and by an analogue of the (35) for the photons: in particular, instead of having γ_r that is dependent only on the r of the point, here we have a γ_e that in general is dependent on all the spatial coordinates of the point. Therefore, for the principle of conservation of energy a particle in motion under the action only of any gravitational field due to a static mass distribution (without electric charge) always conserves the same total energy and consequently the same gravitational mass.

Finally, as in the case due to a spherically symmetric central static mass, we can note that the total energy of a particle in any gravitational field due to a static mass distribution (without electric charge) is always positive.

3 The gravitational field due to an electrically charged rotating mass according to the General Theory of Relativity

3.1 The usual Kerr-Newman solution

The currently used solution (of the Einstein's field equation of the General Theory of Relativity)[24] that represents the curved space-time geometry surrounding an electrically charged rotating mass is the so-called Kerr-Newman solution[25]:

$$ds^2 = \frac{\Delta}{\rho^2}[cdt - a\sin^2\theta\,d\varphi]^2 +$$

---

[24] Cf. EINSTEIN (1916); OHANIAN – RUFFINI pp. 275-293; MISNER – THORNE – WHEELER pp. 383-590; WEINBERG (1972) pp. 151-171; ANDERSON pp. 329-371.

[25] Cf. MISNER – THORNE – WHEELER pp. 875-885.891-894; OHANIAN – RUFFINI pp. 348.355.

$$-\frac{\sin^2\theta}{\rho^2}[(r^2+a^2)d\varphi - acdt]^2 +$$

$$-\frac{\rho^2}{\Delta}dr^2 - \rho^2 d\theta^2 \qquad (154)$$

where ρ^2 and Δ are functions of r and θ:

$$\Delta \equiv r^2 - \frac{2GMr}{c^2} + a^2 + \frac{kGQ^2}{c^4} \qquad (155)$$

$$\rho^2 \equiv r^2 + a^2\cos^2\theta \qquad (156)$$

and M, Q, k, G and a are constants: M is the total mass of the system, Q is the total electric charge of the system, k is the Coulomb's force constant, G is the gravitational constant and a is the angular momentum of the system per unit mass divided by c (that is, $a \equiv \frac{J}{Mc}$, where J is the angular momentum of the system).

Also this metric would imply the presence of an event horizon (for a^2 and Q^2 less than certain values) and of a black hole inside that event horizon. However, we can note that, according to the (154), both the presence of an angular momentum and the presence of a charge Q tend to impede the presence of an event horizon and therefore the presence of a black hole: in fact, if $a^2 > \frac{G^2M^2}{c^4}$ there is not any event horizon and, therefore, there is not any black hole, and analogously if $Q^2 > \frac{GM^2}{k}$ there is not any event horizon and, therefore, there is not any black hole. More in general, if $a^2 + \dfrac{kGQ^2}{c^4} > \dfrac{G^2M^2}{c^4}$ there is not any event horizon and, therefore, there is not any black hole.

But we have seen that the symmetry of the equations of the General Theory of Relativity implies that all the solutions of these equations are symmetric with respect to time and reversible over time, and that, therefore, any solution of the Einstein's field equation (of the General Theory of Relativity) cannot include an event horizon, since an event horizon is evidently not symmetric with respect to time, and consequently cannot include a black hole.

All this implies that the (154) cannot be a true solution of the Einstein's field equation of the General Theory of Relativity.

At this point I report also here my derivation of the correct Kerr-Newman solution[26].

3.2 The correct Kerr-Newman solution

By analogy with the case of the Schwarzschild solution we can extend what we have seen for the case of this solution to other similar solutions of the nonlinear Einstein's equations (or more precisely of the Einstein's field equation) of the General Theory of Relativity[27], in particular to the Kerr solution[28], to the Reissner-Nordstrøm solution[29] and to the Kerr-Newman

---

[26] Cf. PACE (2016b) pp. 52-57.

[27] Cf. EINSTEIN (1916); OHANIAN − RUFFINI pp. 275-293; MISNER − THORNE − WHEELER pp. 383-590; WEINBERG (1972) pp. 151-171; ANDERSON pp. 329-371.

[28] Cf. OHANIAN − RUFFINI pp. 343-346.349-360; MISNER − THORNE − WHEELER pp. 877-878; WEINBERG (1972) pp. 240.349.

[29] Cf. OHANIAN − RUFFINI pp. 346-348.360; MISNER − THORNE − WHEELER pp. 840-841.920-921.

solution[30], which solutions are extensions of the Schwarzschild solution[31].

As for the Kerr-Newman solution[32], that is the solution that represents the curved space-time geometry surrounding an electrically charged rotating mass, we have already seen that the commonly used expression is[33]:

$$ds^2 = \frac{\Delta}{\rho^2}[cdt - a \sin^2 \theta \, d\varphi]^2 +$$

$$- \frac{\sin^2 \theta}{\rho^2}[(r^2 + a^2)d\varphi - acdt]^2 +$$

$$- \frac{\rho^2}{\Delta}dr^2 - \rho^2 d\theta^2 \tag{157}$$

where ρ^2 and Δ are functions of r and θ:

$$\Delta \equiv r^2 - \frac{2GMr}{c^2} + a^2 + \frac{kGQ^2}{c^4} \tag{158}$$

$$\rho^2 \equiv r^2 + a^2 \cos^2 \theta \tag{159}$$

---

[30] Cf. MISNER – THORNE – WHEELER pp. 875-885.891-894; OHANIAN – RUFFINI pp. 348. 355.
[31] Cf. PACE (2016b) pp. 41-57.
[32] Cf. MISNER – THORNE – WHEELER pp. 875-885.891-894; OHANIAN – RUFFINI pp. 348. 355.
[33] Cf. MISNER – THORNE – WHEELER p. 877.

and M, Q, k, G and a are constants: M is the total mass of the system, Q is the total electric charge of the system, k is the Coulomb's force constant, G is the gravitational constant and a is the angular momentum of the system per unit mass divided by c (that is, $a \equiv \dfrac{J}{Mc}$, where J is the angular momentum of the system).

But also here the expression (157) is erroneous: in fact, by means of the fact that the correct Kerr-Newman solution must be equal to the correct Schwarzschild solution for $Q = 0$ and $a = 0$, we can say instead that the correct Kerr-Newman solution should be expressed by this other expression:

$$ds^2 = \frac{\Delta}{\rho^2}[cdt - a\sin^2\theta\,d\varphi]^2 +$$

$$-\frac{\sin^2\theta}{\rho^2}[(r^2 + a^2)d\varphi - acdt]^2 +$$

$$-\frac{\rho^2}{\Delta}dr^2 - \rho^2 d\theta^2 \tag{160}$$

where ρ^2 and Δ are functions of r and θ:

$$\Delta \equiv r^2 + \frac{2GMr}{c^2} + a^2 + \frac{kGQ^2}{c^4} \tag{161}$$

$$\rho^2 \equiv r^2 + a^2\cos^2\theta \tag{162}$$

and M, Q, k, G and a are constants: M is the total mass of the system, Q is the total electric charge of the system, k is the Coulomb's force constant, G is the gravitational constant and a is the angular momentum of the system per unit mass divided by c (that is, $a \equiv \dfrac{J}{Mc}$, where J is the angular momentum of the system).

As we can see, the difference between the correct formula and the incorrect formula is only in the definition of Δ, in which we have respectively $+\dfrac{2GMr}{c^2}$ instead of $-\dfrac{2GMr}{c^2}$.

Moreover, we can see that effectively the correct Kerr-Newman solution (160) is equal to the correct Schwarzschild solution (147) for $Q = 0$ and $a = 0$.

3.3 The consequences of the correct Kerr-Newman solution

Therefore, now we know that the correct Kerr-Newman solution is expressed by this other expression[34]:

$$ds^2 = \frac{\Delta}{\rho^2}[cdt - a\sin^2\theta\, d\varphi]^2 +$$

$$- \frac{\sin^2\theta}{\rho^2}[(r^2 + a^2)d\varphi - acdt]^2 +$$

$$- \frac{\rho^2}{\Delta}dr^2 - \rho^2 d\theta^2 \tag{163}$$

---

[34] Cf. PACE (2016b) pp. 52-53.

where ρ^2 and Δ are functions of r and θ:

$$\Delta \equiv r^2 + \frac{2GMr}{c^2} + a^2 + \frac{kGQ^2}{c^4} \tag{164}$$

$$\rho^2 \equiv r^2 + a^2 \cos^2 \theta \tag{165}$$

and M, Q, k, G and a are constants: M is the total mass of the system, Q is the total electric charge of the system, k is the Coulomb's force constant, G is the gravitational constant and a is the angular momentum of the system per unit mass divided by c (that is, $a \equiv \frac{J}{Mc}$, where J is the angular momentum of the system).

Also in this case, as in that of the correct Schwarzschild solution, we can say that there is not any event horizon and therefore that there is not any black hole[35].

In fact, also here, as in the correct Schwarzschild solution, we have not any singularity in the correct expression (163) for any value of $r > 0$ [36].

Moreover, also here, since the coefficient of $c^2 dt^2$ in the (163) is always not negative, the time (that is, the temporal coordinate) does not become in any case as a spatial coordinate, contrary to the common treatment of the space-time inside the event horizon[37]. Moreover, also here, since the coefficient of dr^2 in the (163) is always not positive, the spatial coordinate r does

---

[35] Cf. PACE (2016b) pp. 36-39.52-53.

[36] Cf. OHANIAN – RUFFINI pp. 324-360; MISNER – THORNE – WHEELER pp. 819-940; PACE (2016b) pp. 36-39.52-53.

[37] Cf. OHANIAN – RUFFINI pp. 324-360; MISNER – THORNE – WHEELER pp. 819-940; PACE (2016b) pp. 38.54.

not become in any case as a temporal coordinate, contrary to the common treatment of the space-time inside the event horizon[38].

Consequently, also in this case the light cones are always orientated in the usual way, in particular there is not any horizontal inclination of the light cones, contrary to the common treatment of the space-time inside the event horizon[39].

Now, according to the General Theory of Relativity, the metric described by the (163) is equivalent to the presence of a gravitational field. Consequently, we could describe this situation as a gravitational field in a flat metric (of the spacetime), with a gravitational potential dependent only on the spatial coordinates, in a similar way to what we have made in the case due to a spherically symmetric central static mass.

Therefore, as in the case due to a spherically symmetric central static mass, for every point there is an escape velocity v_e, that in this case in general is not dependent only on the variable r, but is dependent on all the spatial coordinates. However, in any point this v_e is less than the velocity of light in vacuum c. Moreover, also in this case a photon is always free to go away from the gravitational field.

On the other hand, also in this case, as in the case due to a spherically symmetric central static mass, the gravitational mass is given by the (3), and therefore by an analogue of the (24) for the particles with rest mass greater than zero and by an analogue of the (35) for the photons: in particular instead of having γ_r that is dependent only on r, here we have a γ_e that in general is dependent on all the spatial coordinates. Therefore, for the principle of conservation of energy a particle in motion under the action only of a gravitational field due to an electrically charged

---

[38] Cf. OHANIAN – RUFFINI pp. 324-360; MISNER – THORNE – WHEELER pp. 819-940; PACE (2016b) pp. 38.54.
[39] Cf. OHANIAN – RUFFINI pp. 324-360; MISNER – THORNE – WHEELER pp. 819-940; PACE (2016b) pp. 38-39.54.

rotating mass always conserves the same total energy and consequently the same gravitational mass.

Also here, as in the case due to a spherically symmetric central static mass, we can note that the total energy of a particle in a gravitational field due to an electrically charged rotating mass is always positive.

4 Extension to any gravitational field

Now, we can note that any gravitational field, that is to say also those gravitational fields due to an overall mass distribution that is not static and/or with an electric charge, can be described in every instant as a sum (at the limit an infinite sum) of gravitational fields due each to a portion (also whole) of the mass distribution that would produce a correct Kerr-Newman solution, in general with different centres. In fact the parameters of these correct Kerr-Newman solutions take into account (by means of opportune modifications) in every instant also the interaction between the different parts of the overall mass distribution. Moreover, every correct Kerr-Newman solution does not imply any event horizon and, therefore, does not imply any black hole. Consequently, every part of a correct Kerr-Newman solution does not imply any event horizon and, therefore, does not imply any black hole. Therefore, also a sum (at the limit an infinite sum) of parts of correct Kerr-Newman solutions, in general with different centres, does not imply any event horizon and, therefore, does not imply any black hole. Therefore, any gravitational field, that is to say also those gravitational fields due to an overall mass distribution that is not static and/or with an electric charge, does not imply any event horizon and, therefore, does not imply any black hole.

Moreover, according to the General Theory of Relativity, every gravitational field can be described as a curved metric (of the spacetime), and every curved metric (of the spacetime) is equivalent to the presence of a gravitational field. Consequently, according to the General Theory of Relativity, in general we can describe any gravitational field in a flat metric (of the spacetime), with a gravitational potential dependent in every instant only on the spatial coordinates of every point, in a similar way to what we have made in the case of a gravitational field due to a spherically symmetric central static mass.

Therefore, as in the case due to a spherically symmetric central static mass, in every instant for every point there is a velocity v_e (which, in this case, in general is dependent on all the spatial coordinates of the point) that would be the escape velocity from that point if the gravitational field remained static. However, in any point this v_e is always less than the velocity of light in vacuum c. Moreover, also in this case a photon is always free to go away from the gravitational field.

On the other hand, as in the case due to a spherically symmetric central static mass, the gravitational mass is given by the (3), and therefore by an analogue of the (24) for the particles with rest mass greater than zero and by an analogue of the (35) for the photons: in particular, instead of having γ_r that is dependent only on the r of the point, here in every instant we have a γ_e that in general is dependent on all the spatial coordinates of the point. Moreover, for the principle of conservation of energy a particle in motion under the action only of any gravitational field always conserves the same total energy and consequently the same gravitational mass, if the mass distribution that produces the gravitational field always conserves the same total energy.

Also here, as in the case due to a spherically symmetric central static mass, we can note that the total energy of a particle in any gravitational field is always positive.

As for the case in which the cosmological constant Λ is greater than zero, we have that the presence of the cosmological constant Λ would imply the presence of a repulsive force[40], and consequently also in this case we would always have that there is not any event horizon and, therefore, there is not any black hole. Moreover, also in this case we would have a curved metric (of the spacetime)[41] equivalent to a gravitational field in a flat metric (of the spacetime), with a gravitational potential dependent in every instant only on the spatial coordinates of every point, as in the case with the cosmological constant Λ equal to zero.

Consequently, as in the general case with $\Lambda = 0$, in every instant for every point there is an velocity v_e (which, in this case, in general is dependent on all the spatial coordinates of the point) that would be the velocity with which a particle in that point would be free to go up to $r = +\infty$, if the gravitational field remained static. Moreover, in any point this v_e is always less than the velocity of light in vacuum c. Therefore, also in this case a photon is always free to go up to $r = +\infty$.

On the other hand, as in the general case with $\Lambda = 0$, the gravitational mass is given by the (3), and therefore by an analogue of the (24) for the particles with rest mass greater than zero and by an analogue of the (35) for the photons: in particular, instead of having γ_r that is dependent only on r, here in every instant we have a γ_e that in general is dependent on all the spatial coordinates of every point. Therefore, as in the general case with $\Lambda = 0$, for the principle of conservation of energy a particle in motion under the action only of any gravitational field always conserves the same total energy and consequently the same

---

[40] Cf. OHANIAN – RUFFINI p. 299; WEINBERG (1972) pp. 613-616; PACE (2016b) pp. 59-60.
[41] Cf. OHANIAN – RUFFINI p. 299; WEINBERG (1972) pp. 613-616; PACE (2016b) pp. 59-78.

gravitational mass, if the mass distribution that produces the gravitational field always conserves the same total energy.

Also here, as in the general case with $\Lambda = 0$, we can note that the total energy of a particle in a gravitational field is always positive.

CHAPTER IV
EXPERIMENTAL PROSPECTS

Finally, with regard to the experimental data we can note that so far there is not the experimental certainty of the existence or of the absence of event horizons and black holes[42]. In fact the so-called gravitational waves of a collision between two black holes are been detected only below measurement errors[43].

Moreover, the corrections, that we have proposed here, to the commonly accepted theory are very small in the normal experimental situations, so the fact that so far no difference has been noted between the commonly accepted theory and the experimental results is not strange. In fact, as we have seen, in the usual case of $\frac{2GM}{rc^2} \ll 1$ we have that the difference between the previsions of the erroneous Schwarzschild solution together with the erroneous classical limit and the previsions of the correct Schwarzschild solution together with the correct classical limit is only at the second order in $\frac{2GM}{rc^2}$ [44]. Obviously, we have analogous differences between the previsions of the erroncous Kerr-Newman solution together with the erroneous classical limit and the previsions of the correct Kerr-Newman solution together with the correct classical limit[45]. And all the experiments conducted so far have not had errors so small as to test differences at the second order in $\frac{2GM}{rc^2}$ [46].

---

[42] Cf. GUBSER-PRETORIUS pp. 111-112.132-134.

[43] Cf. GUBSER-PRETORIUS pp. 132-134.

[44] Cf. PACE (2016b) pp. 28-30.35-36.87.

[45] Cf. PACE (2016b) pp. 54-55.

[46] Cf. OHANIAN-RUFFINI pp. 134.142.146; PACE (2016b) p. 87.

Obviously, if we will see a mass M in a sphere of radius smaller than $\dfrac{2GM}{c^2}$ this will not imply automatically the existence of an event horizon (and, therefore, the existence of a black hole), since in my theory these concentrations of mass do not produce an event horizon (and, therefore, do not produce a black hole) in any way. Instead, in this case we should measure the quantities that are different according to the two theories, for seeing which of these two (theories) is the right theory. However, in this case the difference between my theory and the commonly accepted theory is very clear: if that sphere is not bounded by an event horizon, then my theory would be correct and the commonly accepted theory would be incorrect. Vice versa if that sphere is bounded by an event horizon due to gravitational fields, then the commonly accepted theory would be confirmed and my theory would be incorrect[47].

However, recently instruments have been constructed with a sensitivity such as to test these small differences in experiments that are feasible in the solar system. Therefore, it would be appropriate to try to make a crucial experiment that discriminates between the commonly accepted theory and the same theory corrected according to this book[48].

---

[47] Cf. PACE (2016b) pp. 87-88.
[48] Cf. PACE (2016b) p. 87.

GENERAL CONCLUSION:
THERE IS NOT
ANY EVENT HORIZON

We can note that, with the corrections of this book to the common treatment, the motions of the particles in a gravitational field are always symmetric with respect to time and, therefore, reversible over time. In other words, the motion back in time of a particle in a gravitational field is always equivalent to a motion allowed by the correct equations of the gravitational field.

More precisely, the corrections of this book imply that there is not any event horizon, since there is always an escape velocity in every situation and the existence of an event horizon would imply the impossibility of going back for any particle once this particle has gone from the outside of the event horizon to the inside of the same event horizon. In particular, we have seen that the presence of an event horizon would imply the violation of the principle of conservation of energy, since an increase of the kinetic energy, which (increase) would be due to the passage from the outside of the event horizon to the inside of the event horizon, could not counterbalance in any case the decrease of the gravitational potential energy, which (decrease) would due to the passage from the outside of the event horizon to the inside of the event horizon. Consequently, the corrections of this book imply that there is not any black hole, since there is not any black hole without an event horizon. Therefore, this book confutes all the physics that is based on the existence of event horizons and black holes[49].

---

[49] Cf. OHANIAN – RUFFINI pp. 324-381; MISNER – THORNE – WHEELER pp. 819-940; WEINBERG (1972) pp. 207-208.342-349; HAWKING – PENROSE pp.

Moreover, this book, by affirming the impossibility of the existence of event horizons and black holes, permits to explain in a linear way the possibility of the Big Bang, because, according to this book, we cannot have in any case an event horizon or a black hole for the case of the primordial universe[50].

In fact, according to the Big Bang Theory the total energy of the Universe was concentrated in an infinitesimal space and therefore the primordial Universe, according to the common point of view on the General Theory of Relativity, was substantially a supermassive black hole, even though commonly the relativistic physicists refuse to use the name of black hole for the entire Universe, since they know that according to the Big Bang Theory the primordial Universe has not behaved as a black hole.

Now, if the primordial Universe was a black hole not only we would have had the presence of an event horizon surrounding the primordial Universe, contrary to the Big Bang Theory, but also we would have had that, according to the current theory about black holes[51], the primordial Universe would have not expanded in any case: in fact, according to the common theory about black holes[52], inside the event horizon of every black hole all the light cones are orientated towards the centre of the same black hole and therefore there is not any possibility of an

---

11-36.49-75.141-148.152-158; HAWKING (1997) pp. 100-135; PACE (2016b) pp. 83-85.90; THORNE pp. 429-501.

[50] Cf. OHANIAN – RUFFINI pp. 389-437.444-473; MISNER – THORNE – WHEELER pp. 703-799; WEINBERG (1972) pp. 469-597; ANDERSON pp. 444-465; HAWKING – PENROSE pp. 30.37-47.91-122; BARROW (1998) pp. 11-125; DAVIES (2000) pp. 30-46; HAWKING (1997) pp. 51-70.136-165; WEINBERG (1980) pp. 14-165; DAVIES (1997) pp. 133-177.200-214; GOTT pp. 153-197; PACE (2016b) pp. 84-85.

[51] Cf. OHANIAN – RUFFINI pp. 324-360; MISNER – THORNE – WHEELER pp. 819-940.

[52] Cf. OHANIAN – RUFFINI pp. 324-360; MISNER – THORNE – WHEELER pp. 819-940.

expansion of the central mass of a black hole, while the Big Bang Theory affirms that there was a very strong expansion of the primordial Universe[53].

Analogously, my correction to the commonly accepted theory permits the reversibility of a Big Crunch and therefore the possibility of a cyclic Universe[54] with a Big Bang and then a Big Crunch[55] and then a new Big Bang and then a new Big Crunch and so on: in fact, according to this book, all the solutions (of the General Theory of Relativity or of another theory even more correct of the gravitation), and therefore also those that imply a Big Crunch, are theoretically reversible, since are symmetric with respect to time. However, in the hypothesis that the existing Universe is a cyclic Universe, the entropy of the Universe would continue to increase for every cycle Big Bang-Big Crunch[56] and therefore the existing Universe, that has a very small entropy, for thermodynamic reasons should have had origin in a finite past[57].

---

[53] Cf. PACE (2016b) pp. 84-85.

[54] Cf. OHANIAN – RUFFINI pp. 418-428; BARROW (1998) pp. 37-39; DAVIES (2000) pp. 145-150.

[55] Cf. HAWKING – PENROSE pp. 37.44.106-107.117.125; BARROW (1998) pp. 18.37.72; HAWKING (1997) pp. 136.195; DAVIES (2000) pp. 124-131.145-150; WEINBERG (1980) pp. 166-170.

[56] Cf. OHANIAN – RUFFINI p. 422; BARROW (1998) pp. 31-39; DAVIES (2000) pp. 145-150.

[57] Cf. PACE (2016b) p. 85.

BIBLIOGRAPHY

ANDERSON J. L., *Principles of Relativity Physics*, Academic Press, New York 1967

BARROW J. D., *Il mondo dentro il mondo*, Adelphi, Milano 1996³ {Italian translation of BARROW J. D., *The World within the World*, Oxford University Press, Oxford 1988}.

BARROW J. D., *Le origini dell'Universo*, Sansoni, Milano 1998³ {Italian translation of BARROW J. D., *The Origin of the Universe*, The Orion Publishing Group Ltd., London 1994}.

BARROW J. D., *Teorie del tutto: La ricerca della spiegazione ultima*, Adelphi, Milano 2003³ {Italian translation of BARROW J. D., *Theories of Everything: The Quest for Ultimate Explanation*, Oxford University Press, Oxford 1991}.

DAVIES P., *The Edge of Infinity: Where the Universe Came from and how it Will End*, Simon & Schuster, New York 1981.

DAVIES P., *I misteri del tempo*, Oscar Saggi Mondadori, Milano 1997 {Italian translation of DAVIES P., *About Time*, Orion Productions, Adelaide 1995}.

DAVIES P., *Gli ultimi tre minuti*, Superbur Scienza, Milano 2000 {Italian translation of DAVIES P., *The Last Three Minutes*, The Orion Publishing Group Ltd., London 1994}.

EINSTEIN A., *Zur Elektrodynamik bewegter Körper*, in «Annalen der Physik» 17 (1905) 891-921 {English translation: EINSTEIN A., *On the Electrodynamics of Moving Bodies*, in LORENTZ H. A. – EINSTEIN A. – MINKOWSKI H. – WEYL H., *The Principle of Relativity*, Dover Publications, New York 1952, pp. 35-65}.

EINSTEIN A., *Die Grundlage der allgemeinen Relativitätstheorie*, in «Annalen der Physik» 49 (1916) 769-822 {English translation: EINSTEIN A., *The Foundation of the General*

Theory of Relativity, in LORENTZ H. A. – EINSTEIN A. – MINKOWSKI H. – WEYL H., *The Principle of Relativity*, Dover Publications, New York 1952, pp. 109-164}.

GOTT J. R. III, *Viaggiare nel tempo*, Mondadori, Milano 2002 {Italian translation of GOTT J. R. III, *Time Travel in Einstein's Universe*, Houghton Mifflin Harcourt, Boston 2002}.

GUBSER, S. S. – PRETORIUS, F., *I buchi neri*, Bollati Boringhieri, Torino 2018 {Italian translation of GUBSER, S. S. – PRETORIUS, F., *The Little Book of Black Holes*, Princeton University Press, Princeton 2017}.

HAWKING S., *Dal Big Bang ai Buchi Neri*, BUR Supersaggi, Milano 1997[15] {Italian translation of HAWKING S., *A Brief History of Time*, Bantam Dell Publishing Group, New York 1988}.

HAWKING S., *The Universe in a Nutshell*, Bantam Spectra, New York 2001.

HAWKING S. W. – PENROSE R., *La Natura dello Spazio e del Tempo*, Sansoni, Milano 1996[2] {Italian translation of HAWKING S. W. – PENROSE R., *The Nature of Space and Time*, Princeton University Press, Princeton 1996}.

MISNER C. W. – THORNE K. S. – WHEELER J. A., *Gravitation*, W. H. Freeman and Company, New York 1973.

OHANIAN H. C. – RUFFINI R., *Gravitation and Spacetime*, Cambridge University Press, New York 2013[3].

PACE C. M., *All the relativistic temporal paradoxes are completely false*, Youcanprint, Tricase (LE) 2016.

PACE C. M., *There is not any black hole*, Youcanprint, Tricase 2016.

PENROSE R., *La mente nuova dell'imperatore*, BUR Scienza, 2009[5] {Italian translation of PENROSE R., *The Emperor's New Mind*, Oxford University Press, Oxford 1989}.

PENROSE R., *Cycles of Time: An Extraordinary New View of the Universe*, The Bodley Head, London 2010.

REGGE T., *Infinito: Viaggio ai limiti dell'universo*, Oscar Saggi Mondadori, Milano 1996.

SCHWARZSCHILD K., *Über das Gravitationsfeld eines Massenpunktes nach der EINSTEINschen Theorie*, in «Sitzungsberichte der Königlich-Preussischen Akademie der Wissenschaften» 7 (1916) 189-196.

SUSSKIND L., *The Black Hole War*, Little, Brown and Company, New York 2008.

THORNE K. S., *Buchi neri e salti temporali: L'eredità di Einstein*, Castelvecchi editore, Roma 2015 {Italian translation of THORNE K. S., *Black Holes and Time Warps: Einstein's Outrageous Legacy*, W. W. Norton & Company, New York 1994}.

WEINBERG S., *Gravitation and Cosmology: Principles and Applications of the General Theory of Relativity*, John Wiley & Sons, New York 1972.

WEINBERG S., *I primi tre minuti*, Oscar Saggi Mondadori, Milano 1980 {Italian translation of WEINBERG S., *The First Three Minutes: A Modern View of the Origin of the Universe*, Basic Books, New York 1977}.

TABLE OF CONTENTS

Youcanprint
Finito di stampare nel mese di novembre 2019

www.ingramcontent.com/pod-product-compliance
Lightning Source LLC
Chambersburg PA
CBHW080912160726
48000CB00009B/2954